The BUCCANEER

C000083211

Cover artwork of an original Blackburn brochure showing the Buccaneer S.1.

BLACKBURN AIRCRAFT LIMITED
BROUGH · YORKSHIRE · ENGLAND

Blackburn Aircraft Ltd. - A short overview

Robert Blackburn (born 26 March 1885 - died 10 September 1955) started to build aeroplanes in 1909, although his first design, a monoplane, never flew. It was his second monoplane that actually flew when it made its first flight in March 1911. He designed and built more monoplane types, but without great success. In 1913 he built his first floatplane, the Type L and although only one was build, it was impressed by the British Admiralty when the First World War broke out in 1914.

Before the war the Blackburn Aeroplane Co. was founded with a small production workshop at the Balm Road in Leeds, Yorkshire. The company moved later to the nearby Olympia works at the Roundhay Road in Leeds. The monoplane was soon followed by the twin-fuselage T.B. seaplane, which was built in small quantity (8) for the Royal Naval Air Service. The next type that would see limited production was the large twin-engine Kangaroo bomber. After the war, the Kangaroo was also used as a civil airplane.

Eventually all construction work was moved to the Brough factory, where it remained until their last type, the Buccaneer was built. The final company name became Blackburn Aircraft Ltd.

Blackburn also designed and built a special small flying boat racer to participate in the 1923 Schneider Cup flying contests, but this already crashed before the event took place!

Until the outbreak of the Second World War Blackburn produced for the greatest part naval aircraft types, although not all of these were successful. Especially the large flying boat R.B.3 Perth should be mentioned. Although it was produced in very modest numbers, the Perth was extensively used for maritime patrol missions. Blackburn also tried, and mainly failed, to introduce a number of single engine military aircraft types in the late twenties and early thirties. Only the Ripon naval reconnaissance/torpedo bomber biplane was built in some numbers with a total of 96 delivered as a land-based plane with wheels and as a floatplane.

By the late thirties, Blackburn manufactured in somewhat larger numbers the naval aircraft Skua dive bomber and the Roc fighter, fitted with a four-gun turret. Both were a low-wing monoplanes fitted

1 - *The twin-fuselage Type T.B. of 1916 was for its time a very modern design, although the few machines built were hardly used operationally.*

2 - *The Pellet Schneider Cup racer of 1923 already sank before the contest took place.*

3 - *The Lincock was an attempt from Blackburn to enter the biplane fighter market in the late twenties, but at the end only three were built! Also licence production of this type in Italy by Piaggio never emerged and only one single machine was produced. The photo depicts the Lincock III.*

4 - *The Blackburn Ripon of the thirties was built in some numbers as a reconnaissance and torpedo plane for the Royal Navy. Some were exported to Finland. A number was also licence-manufactured in Finland.*

5 - *The Perth was in the thirties a very familiar sight and it served very well although only a handful were actually built.*

6 - *The Blackburn Roc was, just like the Boulton Paul Defiant, a fighter armed with a multi-gun rear turret. It performed definitely less than the Defiant!*

with a Bristol Mercury or Perseus engine, but at the start of the war they were already outdated!

During the Second World War Blackburn produced the Botha, a twin-engine reconnaissance bomber. Although it had very mediocre performances and poor single-engine flying characteristics it was built in fairly large numbers with a total production of 676 aircraft.

Blackburn also developed and built a naval fighter as the B.37 Firebrand, fitted with a Napier Sabre liquid-cooled engine. As a fighter it was never used, but fitted with a Bristol Centaurus radial engine it was used on a small scale as a torpedo bomber but it arrived too late to play any role in the war.

After the war, the most important Blackburn product except for the Buccaneer was the Beverly transport plane, a project taken over from General Aircraft Ltd, where it was known as the G.A.L. 60 Universal freighter. It was built as standard transport plane for the R.A.F., but also here only in limited numbers!

Before the Buccaneer, Blackburn had very little experience with jet aircraft design, but at least they built one, although this is quite unknown. Under their own designation YB-2 Blackburn constructed for Handley Page a small jet-propelled research plane intended as a flying test model for the crescent-winged Handley Page Victor V-bomber. Instead of building this themselves, the project was contracted out initially at Supermarine, but later at General Aircraft. When the latter was taken over by Blackburn, it was completed by Blackburn. This jet plane is better known as the Handley Page HP.88 and it must have contributed almost nothing to the Victor project since it soon crashed after its first flight in 1951 following structural problems. At least this project provided Blackburn some basic knowledge and experience on the construction of fast jet-propelled aircraft!

Concluding we may say that before Blackburn merged into Hawker Siddeley/British Aircraft Corporation it was a relatively small aircraft manufacturer.

The design and final construction of the Blackburn Buccaneer was for this company a very ambitious and challenging project and in fact it was the best aircraft they made during their existence!!

7 - *The Botha was the Blackburn type built in most numbers, although it never excelled as a really good plane (Johan Vischendijk collection).*

8 - *The Blackburn B.20 of 1940 was a very advanced flying boat with a re-tractable hull. Only one was built and it crashed at an early flight stage.*

9 - *The Firebrand torpedo bomber was too late to be used in the Second World War. Some 175 were supplied after the war to the Royal Navy.*

10 - *The Y.B.1 was a naval torpedo bomber and attack plane/U-boat hunter, but it lost from the Fairey Gannet, which showed a remark-*

able resemblance with the Blackburn type. It was powered by an Armstrong Siddeley Double Mamba turboprop engine.

11 - *The ungainly looking, but very effective Blackburn Beverly became after the war a standard R.A.F. transport plane. A few were also used on the civil market.*

12 - *The short-living Handley Page HP.88 was in fact built by Blackburn and gave the company its badly needed experience on jet aircraft design when they started later the Buccaneer project. The single HP.88 flew in military markings with R.A.F. serial number VX330.*

Introduction

The Blackburn Buccaneer was the first jet aircraft specially designed for flying very low under the radar at high subsonic speeds. It was developed in the fifties and entered service at the Royal Navy in 1962. Later it also flew as an attack bomber at the R.A.F. and it even played a role in the Gulf War in 1991 before being retired in 1994 after an operational career that spanned three decades...

The extreme end of the fuselage was used for a split air brake. A quite hidden feature of the NA.39 was its system of boundary layer control to lower the landing speed. The NA.39 had a large rotating bomb-bay below the fuselage that was hydraulically operated. It could be fitted with various weapons and extra fuels. If needed it could carry a tactical nuclear bomb. The new attack plane had a crew of two seated on

Early development and test flying

The need for a special shipboard attack plane as a counter-threat against Soviet warships in the North Sea area became apparent at the early fifties. The idea was to develop an aircraft that could fly during an attack undetected below the ship's radar. The new airplane plane was planned to be jet-powered and fully navalized for operation from aircraft carriers. In 1954 the British Admiralty released specification M.148T for such a type, better known as 'NA.39'. A dozen companies drew up designs, but finally Blackburn received in 1955 the acceptance of their submitted project. Main person responsible for this project was chief designer B.P. Laight. For Blackburn it was a quite challenging project since this company had at that time hardly any experience with jet-propelled aircraft.

The new attack plane was so urgently needed by the Royal Navy that Blackburn received an order for not less than twenty pre-production aircraft, the first one scheduled for its first flight in April 1958. For continuous low-flying the airframe had to be carefully designed to give optimal airflow at the lowest possible drag. As a result the shape of the fuselage was largely dictated by the new 'Area-Ruling' principles as discovered by U.S. NACA aerodynamicist Richard T. Withcomb. In fact, early designs had to be drastically changed to incorporate these new aerodynamic principles. The Whitcomb rules resulted in a very pronounced bulge in the rear fuselage of the NA.39. The final NA.39 design B.103 showed a twin-engine aircraft with the both engines placed in parallel in the centre fuselage section. Each engine had its separate air intake and exhaust. Further the new 'low-flyer' had broad-chord swept wings and a T-tail.

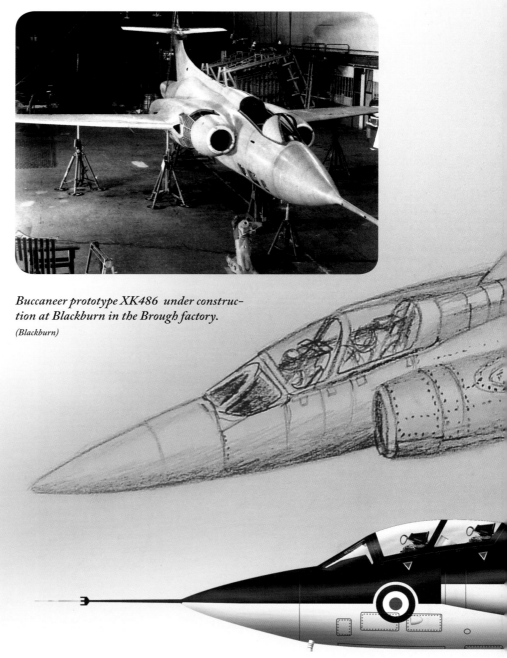

Buccaneer prototype XK486 under construction at Blackburn in the Brough factory.
(Blackburn)

First NA.39 Buccaneer prototype XK486.
(Srecko Bradic)

Martin Baker ejection seats under a large single-piece backwards sliding canopy. Since the new aircraft had to operate at low altitude the structure was strengthened to withstand continuous low-level airflow.

The NA.39 prototype was built under great secrecy. The first flight with the first prototype *XK486* took place on 30 April 1958 from RAE Bedford by Blackburn chief test pilot Derek Whitehead with flight test observer Bernard Watson

in the rear seat. The second prototype *XK487* followed with its first flight on 12 September 1958. The four evaluation aircraft *XK488-XK491* followed over the period November 1958-May 1959.
A further 14 pre-production aircraft for service evaluation with RAF serial numbers *XK523* to *XK536* were delivered by the end of 1961.

NA.39 prototype *XK486* made in September of that year its first official public appearance at the SBAC airshow at Farnborough. By that time it was officially named 'Buccaneer' with S.1 as type designation. The flying program had a serious setback when *XK490* crashed

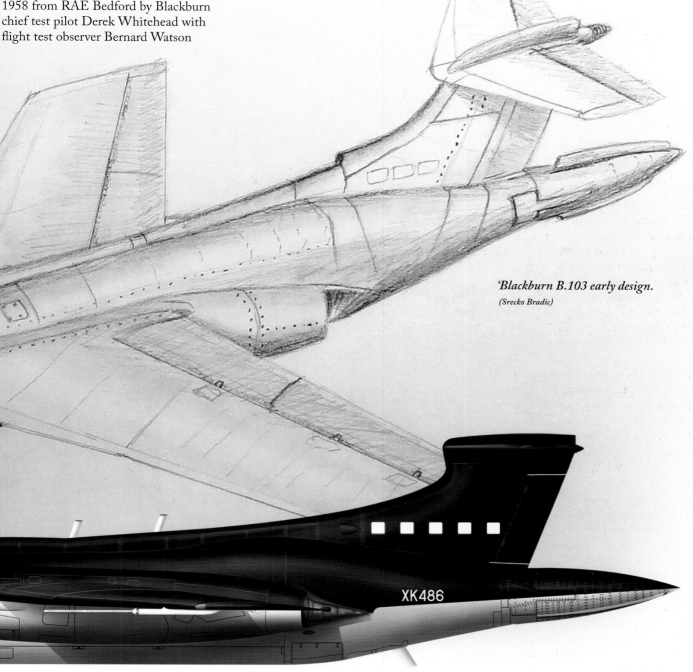

'Blackburn B.103 early design.
(Srecko Bradic)

The first NA.39 Buccaneer prototype XK486, with non-folding wings, at an early test flight in 1958. (Blackburn)

Ground view of the XK489 with a Blackburn Beverley transport plane visible on the left.

(Blackburn)

in October 1959 when flown by a U.S. NASA crew. Pilot W.H. Alford stalled it when he set the thrust lever at zero during landing approach forgetting the blown flaps needed compressed air from the jet engines. The plane was too low for a safe ejection and both Alford and his flight observer J.G. Joyce were killed.

In spite of this, flight testing continued. The first prototypes were in fact nothing more than flying 'empty shells' without any military equipment. The first two machines even lacked folding wings! *XK486* was used for general handling trials.

XK486, XK 488 and XK490 in flight for a display held on 26 March 1959 to celebrate 'Founder's Day' (the birthday of Robert Blackburn). (Blackburn)

NA.39 prototype XK490 with the De Havilland Gyron Junior jet engine.
(Aviodrome collection)

The second prototype *XK487* was used for flutter tests. The third prototype *XK488* was the first to be fitted with folding wings and an arrester hook. Together with pre-production aircraft *XK523* it was used for trials on board of the aircraft carrier HMS *Victorious* in the English Channel in January 1960. *XK490* was the first Buccaneer with the rotating weapons bay for armament trials and an in-flight refuelling probe. *XK491* was used for electrical tests and flight-refuelling sorties with an English *Canberra* bomber converted into a tanker.

- *XK486* was lost on 5 October 1960 when it crashed due to an engine failure. The crew ejected safely.
- *XK487* was later used by Ferranti for radar experiments associated with the B.A.C. TSR.2. It was withdrawn in 1967 and burned a year later.
- *XK488* is now on display in the Fleet Air Air Museum at Yeovilton.
- *XK489* was withdrawn and scrapped in 1964.
- *XK490* crashed with an American crew as already discussed.
- *XK491* was withdrawn and scrapped in 1966 after being used for spinning and ejection seat tests.

In 1960 Blackburn was acquired by the Hawker Siddeley group, being part of British Aircraft Corporation or B.A.C. (later British Aerospace or B.AÈ), and the name of Blackburn as an aircraft manufacturer finally disappeared in 1963. The Buccaneer was the last Blackburn design built.......

In service at the Royal Navy

The Buccaneer S.1

Fourteen Buccaneers S.1 from the pre-production were evaluated by No.700Z Sq. at Lossiemouth. *XK531* and *XK532* were the first to arrive in March 1961 but by the end of this year also *XK533-XK535* arrived, soon followed by the all-white painted *XK535*. They were all extensively test flown. The first operational squadron flying with the Buccaneer was No. 801 at Lossiemouth. They received the first operational production S.1's in July 1962 and soon embarked on board of the aircraft carrier *Ark Royal*. Later they were transferred to the aircraft carrier *Victorious*.

No. 809 squadron was the second one equipped with the Buccaneer S.1, based at Lossiemouth. Later it was re-formed into No. 736 squadron acting as an operational training centre for Buccaneer crews. No.800 Sq. was the third and last squadron to be equipped with the Buccaneer S.1. They were stationed on board of the aircraft carrier *Eagle* and later on the *Ark Royal*.

The total number of Buccaneer S.1 built was 60, including the prototypes and development machines. The last S.1 left the production line in December 1963. Although fully operational, the Buccaneer

No. XN929 was one of the first batch of production S.1 models. We see it here on 8 September 1962 during the exhibition at the Farnborough airshow. The aircraft was painted in anti-radiation gloss white.

(Mick Gladwin collection)

S.1 had as most important shortcoming that its two De Havilland Gyron Junior engines of 3221 kg thrust each provided not enough power for take off at maximum fuel load from an aircraft carrier. To solve this problem temporary a number of S.1's was used as a tanker to fuel-up the other Buccaneers in flight shortly after take off. Final solution was a version with more powerful engines; the Buccaneer S.2. Concluding we can say the Buccaneer S.1 never fully met its expectations and it had a relatively short operational career at the Royal Navy when all machines were permanently grounded after two crashes in December 1970.

Except for a few machines used as museum piece or gate guard and some S.1's being converted into S.2 versions, most were soon scrapped.

Buccaneer no. XK534, one of the batch of evaluation aircraft, at the SBAC Show, Farnborough on 9 September 1961. Also this aircraft was painted in anti-radiation white. The name on the air intake was placed strictly for the airshow! Registration '668/LM' was for Royal Navy 700Z In Flight Testing Unit (IFTU). It was permanently based at Lossiemouth.

XK534 is seen here taxiing to the runway for its demonstration at the 1961 Farnborough airshow.

(Thijs Postma collection)

XK534 taking off at the 1961 Farnborough airshow.
(Thijs Postma collection)

Buccaneer S.1 XK491 used for operational testing. *(Srecko Bradic)*

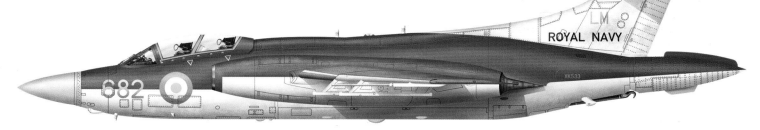

Buccaneer S.1 XK533 no. 682 based at Loss-iemouth *(Srecko Bradic)*

Unguided rocket firing trials with Buccaneer XN981. It was originally supplied as a S.1, but it was later converted into a S.2 when it joined the R.A.F.
(Johan Visschedijk collection)

Buccaneer S.1 XN928 in flight. The 'V' indicated it was stationed on board of the aircraft carrier HMS Victorious.

An all-white Buccaneer S1 XN960 coming in for the landing with the arrester hook out. It was carrying a camera crate in the bomb bay.
(Mick Gladwin collection)

A more detailed shot of XN960 with a full view on the camera crate and with airbrakes fully open. We can clearly see the openings for the flare ejectors.
(Mick Gladwin collection)

Buccaneer S.1 XN969 of RN no 800 squadron on deck of the aircraft carrier R05 HMS Eagle ready for launching.

Buccaneer S.1 XN964 as currently on display outside the Newark Air Museum.

Another shot of XN964 at the Newark Air Museum showing the split airbrakes.

S.1 no. XN967 at Helston Aero Park in Cornwall photographed in June 1985. It was removed ten years later and partially scrapped. The nose section still is in private hands.

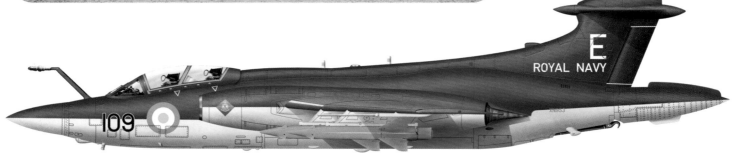

Buccaneer S.1 XN953 no. 109 of Royal Navy Squadron no. 800 based on the aircraft carrier Eagle *(Srecko Bradic)*

Buccaneer S.2 XK526 was the first Rolls Royce Spey powered prototype converted from an S.1 airframe.
(Hawker Siddeley)

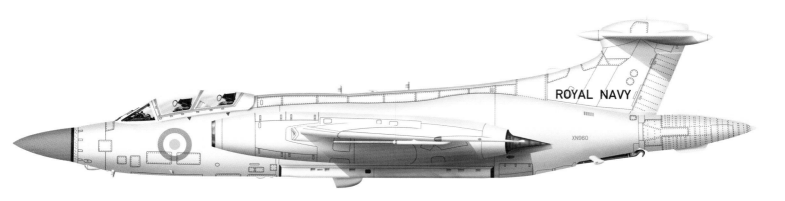

All white Buccaneer S.1 XN960 with camera crate (Srecko Bradic)

Buccaneer S.1 XK488 as photographed in 1985 at the FAA Museum at Yeovilton.
(Photo Nico Braas)

In-flight picture of the XK526 S.2 prototype. It was withdrawn from flying status in 1985 receiving the registration 8684M for a non-flying instructional aircraft. In 1987 it went to R.A.F. base Honington as a gate guard.
(Hawker Siddeley)

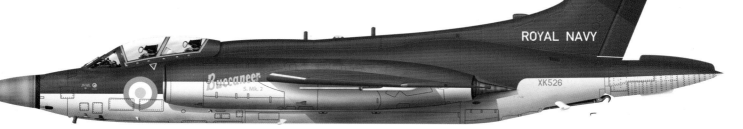

Buccaneer S.2 prototype XK526 (Srecko Bradic)

The Buccaneer S.2

Plans for re-engining the Buccaneer already existed in 1959. As new engine the Rolls Royce Spey was finally selected. It was the same type as used in the De Havilland/B.A.C. Trident airliner. It would fit reasonably well into the Buccaneer airframe without the need for extensive re-designing. The Spey, with designation RB.168-1A was a two-spool bypass engine offering 4990 kg thrust, some 40% more than the Gyron Junior. The Spey was also more fuel efficient than the Gyron Junior giving the Buccaneer a much larger range with the same load of fuel. The most significant change was a new air intake with a much larger diameter. With an improved boundary layer control system and a slightly increased wingspan the first of the two S.2 prototypes, *XK526*, made its first flight on 17 May 1963 by Derek Whitehead. The second S.2 prototype *XK527* followed with its first flight on 19 August 1963. *XK526* was converted to production S.2 standards and was withdrawn from service in 1982. In 1987 it went to Honington as a gate guard.

XK527 was the second S.2 prototype. It was later converted to S.2D standards as shown on this photograph. It carried four Martel missiles.
(Hawker Siddeley)

Buccaneer S.2 XT253 no. 023 of Royal Navy Squadron no. 809 based on the aircraft carrier Ark Royal. (Srecko Bradic)

XK527 was also converted to operational standards and finished its life as an S.2D It was withdrawn in 1991 and scrapped at Lossiemouth in 1993 except for its nose section which is now preserved at New Milton.

The first production S.2 Buccaneer, no. *XN974*, was supplied to the A&AEE for testing in August 1964 and was commissioned on board of HMS *Hermes* in September 1965. In 1971 it went into R.A.F. service and was later converted as an S.2B. In general the S.2 fully met its expectations when compared with the previous version S.1.

In total 133 were produced, including the later S.2B version, in various batches with the last one, no. *XV869* being delivered in December 1968. The Buccaneer was stationed on board of the aircraft carriers *Victorious*, *Hermes*, *Ark Royal* and *Eagle*. Characteristic for these Buccaneers were the tail codes on the tail, being respectively V, H, R and E.

Royal Navy units using the Buccaneer were:
- 700B and 700Z Flight (Lossiemouth) for service trials
- 736 squadron (Lossiemouth)
- 800 squadron (HMS *Eagle*)
- 801 squadron (HMS *Victorious*)
- 803 squadron (HMS *Hermes*)
- 809 squadron (Lossiemouth and HMS *Ark Roya*l)

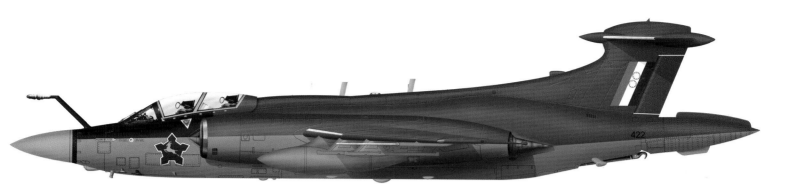

Buccaneer S.50 no. 422 of the South African Air Force. (Srecko Bradic)

It was the intention that the Royal Navy would be equipped with additional aircraft carriers of 50,000 tons, designated as CVA-01. However, the British government decided in 1964 that it did not need any longer aircraft carriers and over the next years all four aircraft carriers were withdrawn from operational use with the *Ark Royal* as last in December 1978. The only new type of aircraft carrier was for the new Sea Harrier and helicopters in the Invincible class. The Royal Navy saw no further need for most of their Buccaneers and only a small number was kept operational, but luckily its days were not over yet as we shall see....

Buccaneer S.2 XT282 from HMS Eagle in a civil hangar with a Cathay Convair CV-880 on the background.

Buccaneer S.2 XT283 of 809 sq grips a cable during a landing on the deck of HMS Ark Royal. Later it would be taken over by R.A.F. 237 OCU squadron as an S.2A.
(Jantinus Mulder slide collection)

The Buccaneer crew of the Royal Navy did their best to look also like a buccaneer! According to the symbol on the air intake cover they are all from RN no. 800 squadron.
(Mick Gladwin collection)

The only operational mission of Royal Navy Buccaneers where bombs were actually dropped took place on 28 March 1967. The oil tanker Torrey Canyon had grounded near Land's End and it was feared that thousands tons of crude oil would be spilled on the Cornish coast. To prevent this, the vessel was bombed by eight Buccaneers from No. 800 and 736 Sq. with in total 42 high-explosive 500 lb bombs. During this strike thirty direct hits were placed. Similar attacks were made on the following days.

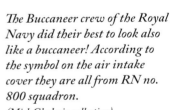

Buccaneer S.2 XV340 from Royal Navy 809 squadron flying over the sea. It flew later as an S.2B at R.A.F. squadrons 16, 12 and 208 and went in 1979 to Brough for fatigue tests. It ended its days at the Pendine firing range as a practice target.

Export Buccaneers

Blackburn/Hawker Siddeley seemed to have with the Buccaneer a very attractive export article. Germany showed a vivid interest in the Buccaneer as a possible replacement for its Hawker SeaHawk and *XK489* was flown for a German delegation at Dunsfold in September 1960. This was continued a year later with a further flight demonstration of *XK534* at Fürstenfeld-bruck in Germany. Eventually all negotiations led to nothing! Also negotiations with India, Australia and Canada led to

Buccaneer S.2 no. XX897 was used as a test-bed for the radar systems of the Panavia Tornado.
(British Aerospace)

HMS Ark Royal based Buccaneer S.2 no. XV343 of no 809 sq. taking off from land. In R.A.F. service as an S.2A it crashed in 1972 near Honington.

nothing although the S.2 was at least test flown by an Indian pilot.

The U.S. Navy was initially also interested in the Buccaneer but already at an early stage they decided to have the U.S. built Grumman A-6 Intruder as low-level attack plane. It must be said that the Intruder was developed using experiences gained in the U.K on the Buccaneer with the Americans having full access to all Buccaneer project data. Just like the Buccaneer the Intruder also had a quite long operational career.

Buccaneer S.2 '035' landing on HMS Eagle, 1971.
(Mick Gladwin collection)

Buccaneer S.2 XT269 was one of the aircraft flown by no. 700Z evaluation squadron at Lossiemouth. In 1972 it rolled off the deck of the aircraft carrier HMS Ark Royal. It was not recovered. (Hawker Siddeley)

The only export customer acquiring the Buccaneer was the Republic of South Africa. Here an order for 16 Buccaneers was placed for a version specially suited to the South African needs as the S.50. The S.50 was roughly based on the S.2, but without the arrester hook and with an additional provision to fit two Bristol Siddeley BS605 rocket motors in the rear fuselage. The folding wings were retained. The South African pilots were trained in the U.K. during the delivery program where the S.50's flew with so-called Class-B civil registrations. Registrations used were *G-2-1* to *G-2-16*. The South African air force registrations were *411* to *426*. Final delivery flight of the first batch to South Africa took place in October 1965 via the Canary Islands. One of these, no. *417*, crashed into the sea some 800 km south of the Canary islands but the crew safely ejected. They were rescued by a Dutch merchant ship.

Buccaneer S.2 no. 238 (serial no. XV336) of Lossiemouth based 801 sq in its element at very low altitude.
(Hawker Siddeley)

Buccaneer S.2 XN978 of 700B evaluation squadron based at Lossiemouth is seen here carrying the photo reconnaissance crate in the bomb bay.
(Mick Gladwin collection)

Buccaneer Photo-Recce crate

By Mick Gladwin

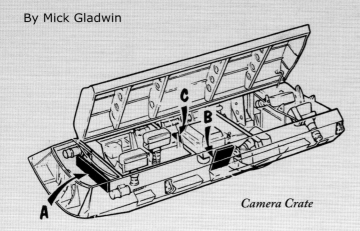

Camera Crate

use the Glow worm 3 inch rocket flares, with eight mounted on the Buccaneer's stores pylons, for night operations. There had been plans to develop a bomb bay tanker pack, but the underwing pack proved adequate.

There was also a plan for a pack with a twin Aden 30 mm cannon, but it was abandoned and the Buccaneer would never carry gun armament. SAAF Buccaneers also had a photographic reconnaissance capability.

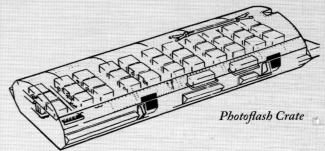

Photoflash Crate

The Buccaneer had a reconnaissance capability, the bomb bay could accommodate a 2,000 litre ferry tank, as well as a photo-reconnaissance "crate" or a cargo container.

The reconnaissance crate could accommodate a photo-flash flare dispenser and up to six F.95 cameras; these could be used in various configurations of long-range and wide angle tasks, that could have the cameras mounted in vertical, oblique, or (using a blister in the pack) forward-looking orientations. The photoflash unit was rarely used, with the Royal Navy preferring to

The RAF inherited some reconnaissance crates along with ex-Royal Navy Buccaneers, but RAF Buccaneers rarely, if ever, flew reconnaissance missions. I once had an RAF Buccaneer crew ask me how to fit one of these crates as they had found one in a hanger, I told them just to leave it alone.

RAF Buccaneer crew on maritime patrols were sometimes issued with hand-held cameras to photograph any possible Warsaw Pact vessels.

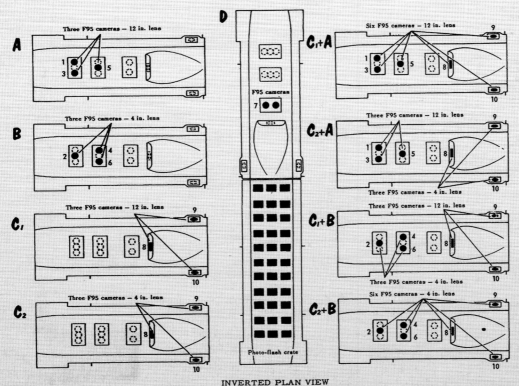

Camera Crate layout.

'HMS Victorious being photobombed by a Blackburn Buccaneer' is the title of this picture. We may conclude it was purposely made to show both Buccaneer and the aircraft carrier! (MoD)

With the growing worldwide anti-Apartheid awareness it was not replaced by a new one! The last S.50's were completed in 1966. They were transported to South Africa by ship. The 15 remaining South African Buccaneers were operational at a single S.A.A.F. squadron, no. 24, based at Waterkloof and flew normally with two underwing fuel tanks and as standard armament four French Nord A.S.30 missiles during strike missions. They were actively used during the 'Border War' with Angola over a period of almost ten years (1978-1988). The South African Buccaneers were retired by the end of June 1991. Just like the Royal Navy Buccaneers, S.A.A.F. Buccaneers also carried out strikes on wrecked tanker vessels to prevent massive spread of crude oil. These strikes took place in 1971 and 1972 on the vessels Wafra and Silver Castle.

Part of the South African Air Force at Waterkloof. This picture shows the relative sizes of the types in service and we can see that the Buccaneer was not much smaller than the Canberra bomber! (S.A.A.F.)

Buccaneer S.50 start from RAF Elvington assisted by two rocket motors. The South African S.50 was the only Buccaneer type fitted with this device.

(Hawker Siddeley)

South African Buccaneer S.50 in the U.K. with the Class-B registration G-2-1 fitted with four AS.30 missiles. It was the first machine supplied and received later its S.A.A.F. registration number '411'. It crashed in January 1974.

(Hawker Siddeley)

Buccaneer S.50 no.422 as displayed at the National Museum of Military History, Saxonwold, Johannesburg, South Africa.

South African nuclear bombers

South Africa was one of the countries having its own nuclear bombs as a deterrent force. In fact it was the only African country having this capability. The 'weapons grade' HEU (Highly Enriched Uranium)235 was manufactured during a secret nuclear programme and it total this yielded enough material to produce a number of nuclear bombs. It was an open secret that Israel has actively supported this program.
The South African Air Force had two military aircraft suitable for the launch of their nuclear weapons: the English Electric Canberra and the Blackburn S.50 Buccaneer.

Eight empty shells intended for the South African nuclear bomb.

Already in 1948 an Atomic Energy Board was established with as final result a nuclear research laboratory near Pretoria in the early sixties. The research also included uranium enrichment. When the international political situation in Africa was such that, just like Israel, the very existence of the 'white' South African republic became under pressure the enrichment of weapons grade uranium was started.
It was the intention to build up a typical 'Cold War' style deterrent force to give a very clear message to neighbouring countries not to start any military adventures against South Africa.
Although preparations were made for a nuclear test explosion in the Kalahari desert in the summer of 1977 this never took place under heavy international pressure.

In spite of this there was enough U235 material enriched to produce a number of nuclear bombs.

These would have been delivered by specially modified versions of the English Electric Canberra and the Blackburn S.50 Buccaneer. However, also nuclear ballistic missiles were at that stage under active development. The South African nuclear bombs were estimated to have a yield of 10 to 18 KT. The bomb had, when fully assembled, a length of 1.8 m and a diameter of 63.5 cm with a weight of 900-1000 kg.

South Africa ended its nuclear weapons programme in 1989. All the bombs (six constructed and one under construction) were dismantled and South Africa acceded to the Treaty on the Non-Proliferation of Nuclear Weapons when South African Ambassador to the United States Harry Schwarz signed the treaty in 1991. On 19 August 1994, after completing its inspection, the International Atomic Energy Agency (IAEA) confirmed that one partially completed and six fully completed nuclear weapons had been dismantled.
Until now, South Africa is the only country in the world that gave up its nuclear armament on a voluntary base!

Cockpit interior of a South African Buccaneer S.50.
(S.A.A.F.)

In service at the R.A.F. ... at last

When the NA.39 was ordered as a new attack plane for the Royal Navy, a land-based version was also offered to the R.A.F. as a replacement for the Canberra bomber. However, at that time they showed no interest since they had selected the B.A.C. TSR-2. It had Mach 2+ capability and could operate both at low and high altitude. Even when the TSR-2 was cancelled in 1965 the R.A.F. still was not interested since they regarded the General Dynamics F-111K as the best alternative. However, when also plans to acquire the

F-111K were cancelled in January 1968 it meant the Canberra still was not replaced! Blackburn/Hawker Siddeley had already earlier submitted a further development of the Buccaneer (Project P.150). It could reach supersonic speeds thanks to a thinner wing and re-heated jet engines. It also had up-dated navigation/attack systems (from the TSR-2) but even after cancellation of the TSR.2 the Blackburn project was not accepted. With the cancellation of the F-111K order, the R.A.F. did not have a suitable modern attack plane!

Reluctantly, an order was placed for 26 Buccaneer S.2B's for the R.A.F. (*XW525-XW550*). In addition the R.A.F. agreed to take over a number of Royal Navy S.2's after some minor modification designating these as the S.2A. The first S.2A delivered was *XV350* which made its first flight in this modified form on 11 February 1969. Later on another twenty S.2B's were ordered for the R.A.F. with another three supplied for weapon trials at R.A.E. Bedford. The last S.2B was delivered in January 1977.

Buccaneer S.2B XV341 of R.A.F. no.15 Squadron. XV341 was originally an S.2 Royal Navy machine. When it went into R.A..F. service it was converted to S.2B standard. In 1985 it crashed at Lossiemouth.
(Mick Gladwin collection)

Below:
XX891 also was a Buccaneer S.2B from R.A.F. no.15 Squadron. In 1983 it crashed at Laarbruch, Germany.
(Mick Gladwin collection)

Buccaneer S.2B XV352 on 25.05.1991 at Mildenhall during an Open Day. It was one of the Buccaneers taking part in the Gulf War and was proudly shown still in 'war colours'. It carried the name 'Tamdu' with tail code 'U'.

(Photo Wim Zwakhals)

A Buccaneer S.2 of R.A.F. no.208 Sq with a camera crate in the bomb bay. These crates were rarely used by the R.A.F. Buccaneers!

(Mick Gladwin collection)

Buccaneer XX901 of R.A.F. no. 208 sq in its early three-tone colour scheme. It was later used in the Gulf War as 'Kathryn, the Flying Mermaid' with tail code 'N'

(Mick Gladwin collection)

The S.2B could be easily distinguished by a bulge below the fuselage to accommodate an enlarged fuel tank of 1932 l (425 UK gal.) in the bomb bay. Other changes included up-to-date R.A.F. avionics and a strengthened undercarriage. It had a higher all-up weight of 28,123 kg. Although the catapult hold-back was removed the arrester hook and folding wings were retained to avoid additional re-design work. Of course the arrester hook was not used, but the R.A.F. Buccaneers regularly used the wing folding mechanism for efficient parking.

The following R.A.F. squadrons were equipped with the Buccaneer:

- No. 12 Squadron (Honington)
- No. 15 Squadron (Laarbruch, Germany)
- No. 16 Squadron (Laarbruch, Germany)
- No. 208 Squadron (Honington and Lossiemouth)
- No. 216 Squadron (Honington and Lossiemouth)
- No. 237 OCU at Lossiemouth/ Honington used the S.2A and S.2B for conversion training.

No.15 and 16 squadron at Laarbruch were based not very far from the East German border and since the R.A.F. Buccaneers could be used to drop a tactical nuclear load two Buccaneers were on permanent readiness in a special hardened aircraft shelter. They could be in the air within 15 minutes armed with a WE.177 nuclear bomb.

Buccaneer S.2B no. XV353 of no.208 sq photographed during a visit at Leeuwarden, the Netherlands on 19 June 1990.

(Photo Minze Veenstra)

Detail of the fully opened split airbrake on a Royal Navy Buccaneer.

(Wikimedia Public Domain)

Buccaneer S.2B XW530 on display as an eye-catcher at a fuel station near Lossiemouth. The photo was taken in August 1999.

(Photo Minze Veenstra)

Buccaneer XW528 in a factory-fresh colour
scheme still without any squadron markings
before it went to R.A.F. 237 OCU.
(Johan Visschedijk collection)

Buccaneer S.2B XN981 of R.A.F. no.12 sq
photographed on 4 June 1987 at Twenthe,
the Netherlands during a visit.
(Photo Minze Veenstra)

Buccaneer S.2A XV154 of 237 OCU.
(Johan Visschedijk colection)

In August 1977 R.A.F. Buccaneers participated for the first time in the Red Flag combat exercises at Nellis AFB in the United States. One of the results was that the R.A.F. Buccaneers were made suitable to carry a pair of AIM-9 Sidewinder missiles for self-defence. In practice the very low flying Buccaneers proved to be more than a match for some of the most modern fighters when flying at their usual height of 30-180 m.....

Metal fatigue problems:
Although the airframe of the Buccaneer was immensely strong, the continuous flying over land created much more stress on the airframe. Conditions over sea were mostly quite turbulence-free, which was definitely not the case over land. The first sign of a fatigue problem occurred on 7 February 1980 when a Buccaneer wing folded in flight. Initially the cause was believed to be a faulty wing bolt, but when another Buccaneer broke up in the air during a Red Flag exercise fatigue cracks were discovered in the wing and all Buccaneers were grounded. The aircraft that could not be economically repaired were withdrawn and scrapped; the rest of the Buccaneer fleet was extensively strengthened. In total some 60 Buccaneers were returned in full operation.

Buccaneer S.2B XW533 of no. 237 OCU photographed in the early 70's at Biggin Hill.
(*Jantinus Mulder slide collection*)

Buccaneer S.2B XN976 of no. 208 sq in a special colour scheme to celebrate the squadron's 75th anniversary. The Sphinx on the tail marked the squadron's association with the Middle East!
(*Jantinus Mulder slide collection*)

Buccaneer S.2B XV360 of R.A.F. 237 OCU is showing its empty bomb bay (Sep 1974).
(Jantinus Mulder slide collection)

Buccaneer S.Mk.2B XW529 was used by MoD(PE) for weapons trials. It carries a Sea Eagle anti-shipping missile under its wing.
(British Aerospace)

Martel Missile (AJ.168) (AS.37)

As standard armament for attack of enemy targets the Buccaneer carried under its wings four Martel air-to-surface missiles. The Martel was developed jointly by the Anglo-French companies Hawker-Siddeley and Matra. It went into operational service at both countries. The French Armee de l'Air used their Mirage III as a launching platform.

Basically there were two versions of the Martel. The missile designated as AS.37 was fitted with a device homing on the radar installation of enemy ships and anti-aircraft batteries.

A different version designated as AJ.168 had for homing to its target a nose-mounted TV camera. Externally the two versions could be easily distinguished by the shape of their nose cone.

The Martel (an abbreviation for Missile-Anti-Radiation-TELevision) was a weapon with a length of 4.18 m, a diameter of 0.4 m and a weight of 550 kg. It was powered by a two-stage rocket motor with solid propellant and was fitted with cruciform small wings with a span of 1.2 m and had a maximum range of 60 km. Once fired it cruised to its target with a speed of some 1000 km/h (Mach 0.9). It was armed with an armour-piercing explosive load of 150 kg with a proximity fuse. The Buccaneer used the Martel until it was replaced by the Sea Eagle.

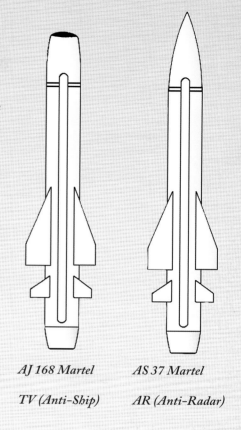

AJ 168 Martel

TV (Anti-Ship)

AS 37 Martel

AR (Anti-Radar)

Sea Eagle

For attacks on enemy ships British Aerospace ('BAe Dynamics') developed a missile more or less in the same class as the French Exocet. The weapon had a length of 4.24 m, cruciform wings with a span of 1.2 m, a body diameter of 0.4 m and a total weight of 580 kg. It carried an explosive load of 230 kg PBX (Polymer-Bound-Explosive) and was semi-armour piercing. It was powered by a small paraffin-burning turbojet and was inertial-guided with active radar homing. Just like the French Exocet it was a sea-skimming weapon, and as such much more difficult to intercept than the Martel. Its load was regarded as sufficient to sink or disable a large military vessel like an aircraft carrier and its systems were protected against jamming or decoys.

BAE SYSTEMS

It was operated both by Royal Navy and R.A.F. and it was also exported to Saudi Arabia and India. Just like the Martel it flew at subsonic speeds (up to Mach 0.9). However, it had a much larger range than the Martel: more than 110 km with a flight duration of almost seven minutes.

Sea Eagle was first fired in 1981. Development of a projected Mk-2 was abandoned.

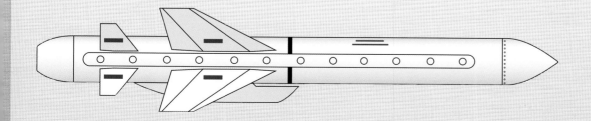

Buccaneer S.2A XV350 with four Martel missiles under its wings.
(Hawker Siddeley)

Publicity shot of Buccaneers of no.12 squadron showing XW537 at the foreground accompanied by a second Buccaneer. The squadron was based at Honington.
(Mick Gladwin collection)

Above & right: RAF no.12 sq Buccaneer XV869 in October 1987 at Twenthe AFB during a squadron rotation with KLu 313 Squadron. The sticker on the tail is from KLu 313 sq!
(Photo Johan Visschedijk)

Buccaneer S.2B XW550 of R.A.F. no.16 sq photographed in December 1978 at Twenthe during a squadron rotation.
(Photo Johan Visschedijk)

Buccaneers of no. 208 sq flying low over the pyramids in Egypt.
(Mick Gladwin collection)

A typical publicity picture of a mixed formation of Royal Navy and R.A.F. Buccaneers.
(Mick Gladwin collection)

WE177 nuclear bomb

Replacing the earlier generation of much heavier nuclear air-launched bombs, Hunting Engineering at Luton developed and built in the early sixties a suitable nuclear weapon that could be launched by relatively small aircraft instead of with the much larger V-bombers.

It was specially developed to be carried by the Blackburn Buccaneer and B.A.C. TSR-2 tactical attack bombers. When the TSR-2 was cancelled the Tornado was later also used as an alternative.

Loaded with weapons-grade Uranium235 the bomb had a maximum yield of some 10 kiloton TNT. Basically this was more than sufficient for the total destruction of its main targets: large enemy vessels, complete harbour installations, military production facilities and large military bases.

Designated as 'WE177A' it became operational in the U.K. by 1966.

The WE177A was basically a streamlined casing with four small stabilizing fins fitted with its nuclear load with the following characteristics:

Yield:	variable form 0.5 to 10 KT
Weight:	272 kg
Length:	2.85 m

A WE177A instructional bomb without its nuclear load is displayed in the Science Museum at London.

The estimated number of bombs manufactured is 107. It was armed before the start of a nuclear mission with two special keys.

It had a pre-determined setting for detonation at 40 m and after launch it was slowed down with a small parachute. The Buccaneer used a standard LABS or toss bombing technique where the weapon was released at full speed on the top of a climb. From here it followed a ballistic trajectory to its target. After release the Buccaneer made a half-loop and a roll to fly away as fast as possible into the opposite direction of the detonation point.

The WE177A was operational until 1992.

A somewhat larger and heavier version carrying a 450 KT thermonuclear load was known as the WE177B. It was a typical weapon for strategic use by the V-bombers.

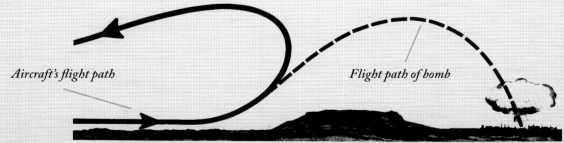

Aircraft's flight path

Flight path of bomb

This drawing from an early Buccaneer brochure in the German language shows the principle of LABS/toss bombing by the Buccaneer.

The Buccaneer goes to war

Lebanon crisis 1983
- Operation Pulsator:

The first time the Buccaneer saw action was in the Lebanon crisis of late 1983 when Druze armed forces threatened to overthrow the existing government. UN forces were sent in to prevent things going out of control into a civil war. These forces included 100 British troops. To give their soldiers air support six Buccaneers from no. 12 and 208 squadron were based at Cyprus under the mission name Operation Pulsar. They were armed with laser guided bombs and a Pave-Spike laser for target illumination. The aircraft were also fitted with chaff to disturb SAM radar systems. Eventually no bombs were dropped and the Buccaneers only let their presence known by making low-pass flights over Beirut.

Gulf War 1991 - Desert Storm:

In the early nineties the Buccaneer was in fact already scheduled for final retirement. However, things went different when Iraq occupied Kuwait with as result a military intervention known as Operations Desert Shield and Desert Storm. Although

Buccaneer S.2B 'Dirty Harriet' in full Gulf War colours flying near Lossiemouth. It was kept there on reserve during the Gulf War. (MoD)

Buccaneer S.2B XW547 on display at the Cosford Aerospace Museum in its Gulf War colour scheme. XW547 carried the name and pin-up artwork of 'Pauline, the Guiness Girl' with tail code 'R'.

the U.K. Ministry of Defence did not show initially any interest to employ the Buccaneer in this conflict, it was the only operational aircraft that could be used for providing target laser illumination for the new Paveway II laser-guided precision bomb. Then, on 23 January 1991 came the order to prepare a dozen Buccaneers for active duty in the Gulf war as target laser illuminators for the Tornado bombers. For this Middle East detachment the best aircraft were selected from R.A.F. no. 12 and 208 squadrons and from no. 237 OCU. In the shortest possible time, a new radio was fitted (Hawk Quick II secure radio's), a new IFF (Identification-Friend-or-Foe) system, a pair of AIM-9L Sidewinder missiles for self-defence and laser equipment for the Paweway II precision bombs. They were also sprayed in a desert scheme to reduce visibility (colour 'Desert Pink'). On the vertical tails single-letter identifications were placed for each Buccaneer; respectively S, L, O, G, U, N, R, E, A, P, I and T. It was no coincidence that these letters formed the words SLOGUN and REAPIT!

Except for the camouflage, most of these Buccaneers were also fitted with WW-II style pin-up paintings on the right fuselage side below the cockpit. These paintings were sponsored by a number of Whisky brands. Also, the name of the company was painted near the pin-up.
The left side showed a black pirate's flag and the word 'Sky Pirates'.
The following Buccaneers took part in the Gulf war:

Pin-up artwork below the cockpit of Buccaneer S.2B XX885 'Hello Sailor, Caroline', sponsored by Famous Grouse. The plane is now being restored as G-HHAA into airworthy condition.

(Jantinus Mulder slide collection)

Buccaneers in the Gulf war:

Aircraft Code	Tail code	Nose art name	Whisky mark
XV352	U	Tandu	-
XW530	E	-	Glenmorangie
XW533	A	Fiona, miss Jolly Roger	Glenfarclas
XW547	R	Pauline, the Guiness girl	The Macollan
XV863	S	Debbie, Sea Witch	Tamnavoulin
XX885	L	Caroline, Hello Sailor	Famous Grouse
XX889	T	-	Longmom
XX892	I	-	Glen Lossie
XX894	O	-	Aberlour
XX895	G	Lynn, the Jaws	Glenfiddish
XX899	P	Laser Lips Laura	Linkwood
XX901	N	Kathryn, the Flying Mermaid	Glen Elgin

Buccaneer XV332 with nose art 'Dirty Harriet' was held in reserve at Lossiemouth; it missed the name of a Whisky mark on its nose.

All 12 Buccaneers selected were flown to its Middle East base Muharraq, Bahrain. The first two Buccaneers made their ferry flight already 3 days after receiving the order! Here, an intensive training period followed in close cooperation with the Tornado's they had to escort. The pilots flying these machines had never used the precision bombs previously!

The first combined mission took place on 2 February 1991 when two Buccaneers (+ one further back-up aircraft) escorted four Tornado's during an attack on a road bridge across the Euphrate river'. The mission was a complete success. More missions followed with this combination of Buccaneers and Tornado's; five or six attacks a day became routine! When it was evident that the Buccaneers were not any longer under threat of enemy aircraft, the Sidewinders were deleted and changed for laser-guided bombs. Over the next period of the war, also Buccaneers dropped their precision bombs with great success most of the times.

Altogether the 12 Buccaneers flew 250 sorties during Operation Desert Storm without losing a single aircraft; a great achievement for a type to be phased out! The last Buccaneer mission was flown on 27 February 1991. After spiking targets at Shayka Mazar airfield for the Tornado's, the lead Buccaneer flown by Flt. Lt. Glenn Mason and Sqn. Ldr. Norman Brown released their own Paveway II precision bomb after they had laser-targeted ('spiked') it on a taxiing Antonov

Buccaneer S.2B, XX894 of no. 12sq with tail code 'O' on exhibit on 1 September 1991 at Kleine Brogel in Belgium.
(Photo Minze Veenstra)

Buccaneer S.2B XX889 of no. 208 sq in Gulf War desert scheme with tail code 'T' and 'Sky Pirates' emblem on its fuselage. The other side of the fuselage showed only the Whisky name 'Longmorn'.
(Jantinus Mulder slide collection)

An-12 transport plane. The bomb did not detonate, but the kinetic energy released at impact was sufficient to put this plane definitely out of use! The silhouette of the plane was proudly painted as a killer-mark under the cockpit of XX885 Caroline-Hello Sailor they had used. This mission ended quite literally with a loud bang when the number two Buccaneer, flown by Wg Cdr Bill Cope and Flt Lt Carl Wilson, also managed to make a direct hit with their laser-guided bomb. This time the LGB did explode resulting in complete destruction of a C-130 Hercules captured earlier by the Iraqi in Kuwait. The Buccaneers had dropped in total 48 LGB's resulting in the destruction of 24 bridges and 15 other military targets (including the two transport planes!).

On 17 March 1991 the mission for the Buccaneer was ended. They flew back non-stop to Lossiemouth. In spite of being warmly welcomed by all parties and media, it was the beginning of the end..

Buccaneer S.2B XV863 of no. 208 Sq was also used in the Gulf war with pin-up 'Debbie, Sea Witch' on the fuselage and tail marking 'S'. Mission details are painted below the cockpit! (Jantinus Mulder slide collection)

Another Gulf War veteran: XX899 'Laser Lips Laura' with tail code 'P' photographed in 1991 at Lossiemouth.

(Photo Wim Zwakhals)

Museum piece

Buccaneer S.2B XV865 as currently on display at Duxford in early no. 208 sq colours.
On the ground we see the Martel version fitted with a TV guidance camera in the nose.
(Photo Edwin Hoogschagen)

XV965 at Duxford showing wing folding details. We can also see the tail end of a TV-guided Martel missile.
(Photo Edwin Hoogschagen)

Tail-end of XV965 at Duxford.
(Photo Edwin Hoogschagen)

RAF Buccaneer S.2B XX900 of no. 12 sq. It is now owned by the British Aviation Heritage Bruntingthorpe, Leicestershire in taxiable condition.
(*Jantinus Mulder slide collection*)

Final phase out and retirement

When the Gulf war started, only the operational R.A.F. squadrons, nos. 12 and 208 and no. 237 OCU still had Buccaneers on their strength and many of the 60 still in service after the metal fatigue problem was solved were by that time already retired. The few remaining Royal Navy Buccaneers were retired some years earlier and replaced by the Sea Harrier (which is now also phased out!).
In spite of the successes in the Gulf War, the Buccaneer was withdrawn from service and replaced with the Tornado. By mid-1993 no. 208 Squadron had become the only operational user of the type. The last Buccaneers were withdrawn in March 1994 when 208 Squadron was disbanded. The three Buccaneers from the Ministry of Defence flying at the R.A.E. for weapons were used until 1985. They were: *XV344*, *XW987* and *XW988*.

XV344 went to Farnborough, where it still is on static display.
XW987 and *XW988* were withdrawn from flying status in 1985 and acquired by aviation enthusiast Mike Beachy Head in South Africa. Both were still in airworthy conditions and received in South Africa the civil registration *ZU-BCR* and *ZU-AVI*. They were operated by Thunder City in Cape Town, together with Buccaneer S.2B *XW986* (*ZU-NIP*) in the airshow circuit, but in 2011 all three machines were offered for sale when Thunder City closed its gates because of the bad economic prospects. *ZU-NIP* is currently owned by Ian Pringle. It is still in airworthy condition as has flown at an airshow in February 2012 together with *ZU-BCR*! All three Buccaneers are still offered for sale!

Versions and production list
The first three NA.39 prototypes *XK486-XK488* were purely intended for aero-dynamic research and were in fact 'flying shells' without any military equipment. *XK489* and *XK490* were already navalized with folding wings and an arrester hook. Further they carried provisions for future armament testing and a modified nose for radar testing. *XK491* and the first NA.39 pre-production aircraft were fitted with autopilot, and an enlarged nose for the final operational radar system

All Buccaneers were produced at the Blackburn production facilities at Brough, near Kingston-upon-Hill.

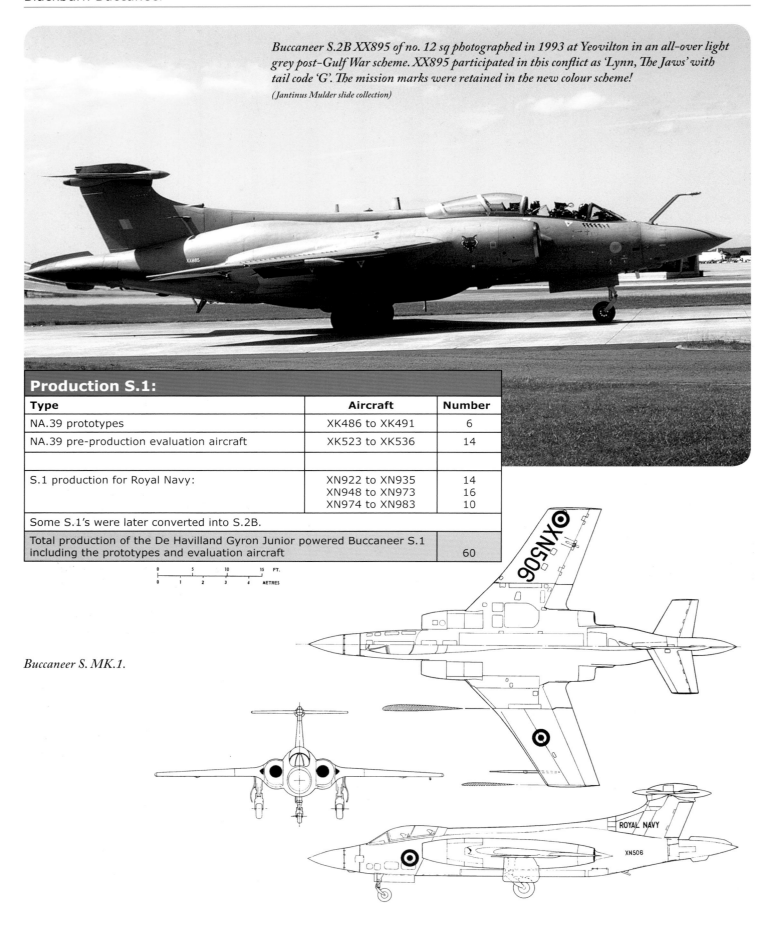

Buccaneer S.2B XX895 of no. 12 sq photographed in 1993 at Yeovilton in an all-over light grey post-Gulf War scheme. XX895 participated in this conflict as 'Lynn, The Jaws' with tail code 'G'. The mission marks were retained in the new colour scheme!

(Jantinus Mulder slide collection)

Production S.1:

Type	Aircraft	Number
NA.39 prototypes	XK486 to XK491	6
NA.39 pre-production evaluation aircraft	XK523 to XK536	14
S.1 production for Royal Navy:	XN922 to XN935	14
	XN948 to XN973	16
	XN974 to XN983	10
Some S.1's were later converted into S.2B.		
Total production of the De Havilland Gyron Junior powered Buccaneer S.1 including the prototypes and evaluation aircraft		60

Buccaneer S. MK.1.

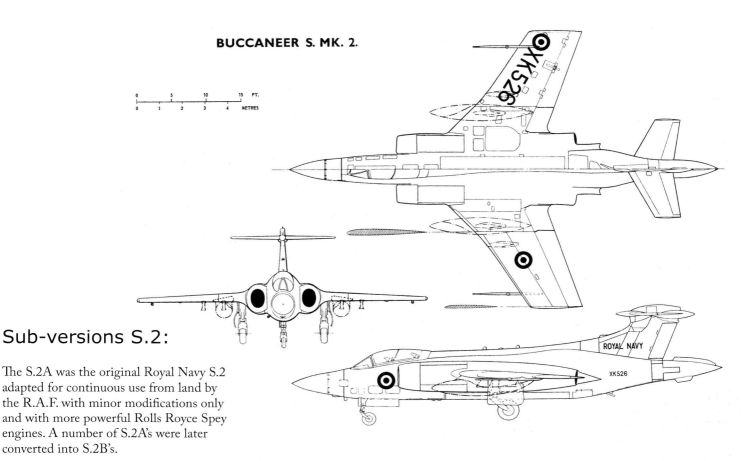

BUCCANEER S. MK. 2.

Sub-versions S.2:

The S.2A was the original Royal Navy S.2 adapted for continuous use from land by the R.A.F. with minor modifications only and with more powerful Rolls Royce Spey engines. A number of S.2A's were later converted into S.2B's.

Production S.2:		
Type	**Aircraft**	**Number**
NA.39 S.2 evaluation aircraft with R.R. Spey engines:	XK526 and XK527	2
S.2 Royal Navy production models:	XN974 to XN983 [1]	10
	XT269 to XT288	20
	XV152 to XV168	17
	XV332 to XV361	30
	XV863 to XV869	7
8 cancelled	XV870 to XV879	
Total		84
S.2B R.A.F. production models:	XW525 to XW550	26
	XX885 to XX901	17
	XZ430 to XZ432	3
- for service trials at the R.A.E	XW986 to XW988	3
Total		49
S.50 S.A.A.F	411 to 426	16
Total production of Rolls Royce Spey powered Buccaneer S.2/S50		151
Total production number Buccaneer all versions		211

(1)   originally planned as S.1

The S.2B was the version exclusively built for the R.A.F. It had a somewhat higher all-up weight, a strengthened undercarriage and an enlarged bomb bay, making a distinctive bulge under the fuselage. It could carry additional fuel in this enlarged bomb bay. Although the catapult hold-back was removed, it still had an arrester hook and folding wings. Most machines could be fitted with four underwing Martel missiles. In total the R.A.F. received 46 machines that were built and delivered between 1973 and 1977. Another three (*XZ430* to *XZ432*) were supplied to the Ministry of Defence for various weapon trials.

The S.2C was a small batch of eight S.2B aircraft modified back to S.2A standards for use by the Fleet Air Arm.

The S.2D was basically a small batch of nine S.2B for the Fleet Air Arm adapted to be equipped with the Martel.

Accidents and incidents:

With its long operational career it was inevitable that Buccaneers would be lost in accidents. The following are known:

Buccaneer S.1:

-**XK486**   crashed on 5 October 1960 during a test flight at Little Weighton after engine failure. The crew safely ejected.

-**XK490**   prototype crashed on 12 October 1959 at Lindhurst when flown by a U.S. NASA crew when it stalled after engine shut down. Both occupants were killed.

-**XK524**   crashed due to tailplane stall on 13 May 1965. The crew safely ejected after the use of explosive cord in the canopy (actually the first use of this device).

-**XK528**   crashed on 30 June during weapons trials at Luce Bay.

-**XK529**   crashed on 31 August 1961 during catapult trials on HMS *Hermes*. The crew was killed.

-**XK535**   was damaged beyond repair on 18 August 1962 at Lossiemouth during the landing approach. The damaged plane was used on the Shoeburyness firing range as target.

-**XN922**   was damaged beyond repair on 5 July 1962 at Boscombe Down.

-**XN927**   crashed on 25 March 1964 near Elgin.

-**XN948**   ditched in the sea on 26 November 1964 near HMS *Victorious* at Changi, Singapore.

-**XN949**   crashed on 8 August 1965 at Moray Firth.

-**XN950**   crashed on 28 March 1966 after an overshoot at Lossiemouth.

-**XN951**   crashed on 1 December 1970 after an overshoot at Lossiemouth.

-**XN952**   crashed on 9 February 1963 at Holme-on-Spalding Moor (HOSM).

-**XN954**   was purposely dumped off HMS *Ark Royal* on 14 April 1974 for a Royal Navy safety film.

-**XN958** crashed on 20 November 1965 near Singapore.

-**XN961**   crashed on 25 July 1965 at Helmsdale. The crew was killed.

-**XN966**   crashed on 24 January 1964 on the runway of Lossiemouth.

-**XN968**   crashed on 8 December 1971 in a forest near Lossiemouth.

-**XN969**   crashed on 9 October 1963 near Khormaksar, Yemen.

Buccaneer S.2B XW988 was one of the three Buccaneers used by the R.A.E. for weapons testing in a highly visible yellow/black colour scheme. It was photographed at West Freugh. (Jantinus Mulder slide collection)

Buccaneer S.2 XW988 as used by the R.A.E. for weapons testing. (Srecko Bradic)

-*XN970* crashed on 25 March 1966 at Beira, Mozambique while operating from HMS *Eagle.*

-*XN975* crashed on 14 July 1978 near Bruggen after a near miss with a German helicopter.

-*XN979* crashed on 9 June 1966 near Lizzard Point; cockpit section still exists.

Buccaneer S.2:

-*XN977* (S.2B) was damaged after a heavy landing on 8 March 1982 and never repaired.

-*XN97*8 (S.2B) crashed on 5 June 1971 near Creil, France.

-*XN980* (S.2) crashed on 3 March 1969 after collision with Buccaneer *XV159* near Wick.

-*XT269* (S.2) rolled off deck of HMS *Ark Royal* on 15 February 1972 and was not recovered.

-*XT276* (S.2B) had a Cat. 3 accident at Laarbruch, Germany on 28 August 1980 and was not repaired.

-*XT282* (S.2) crashed after an hydraulic failure on 31 August 1970 near Moray Firth.

-*XT285* (S.2A) crashed on 5 July 1978 near West Freugh.

-*XV153* (S.2) crashed on 6 October 1966 during evaluation flight on board of HMS *Victorious*.

-*XV158* (S.2) ditched on 20 May 1968 in Moray Firth following hydraulic failure.

-*XV159* (S.2) crashed on 3 March 1969 in the North Sea after collision with *XN980*.

-*XV160* (S.2B) crashed on 20 September 1986 after it stalled during an attack manoeuvre near Sardinia.

Buccaneer S.2B XV163 of 237 OCU taxiing with its wings still folded.
(Jantinus Mulder slide collection)

-*XV162* (S.2B) crashed on 13 June 1972 in the North Sea near Bridlington.

-*XV164* (S.2) crashed on 16 September 1969 at Beinn Ruadh near Strathy Point.

-*XV166* (S.2B) crashed on 4 March 1976 during landing approach on Honington.

-*XV167* (S.2A) crashed on 29 January 1970 during catapult launch from HMS *Hermes*.

-*XV335* (S.2A) crashed on 1 July 1968 at North Minch.

-*XV339* (S.2A) crashed on 6 October 1972 during a test flight from RNAY *Sydenham*.

-*XV341* (S.2B) crashed on 14 July 1985 on landing at Lossiemouth.

-*XV343* (S.2A) crashed on 12 April 1973 near Honington.

-*XV345* (S.2B) crashed on 7 February 1980 at Nellis AFB, U.S.A. following fatigue failure.

-*XV346* (S.2) ditched on 13 February 1969 near Tarbar Ness.

-*XV347* (S.2) was destroyed on 9 December 1971 after an engine fire during taxiing at Lossiemouth.

-*XV348* (S.2B) crashed on 31 October 1977 after hitting power cables at Glomford, Norway.

-*XV351* (S.2D) crashed on 11 November 1974 near Wash.

-*XV360* (S.2A) crashed on 29 July 1978 near Covehithe.

-*XV867* (S.2B) was scrapped on site after an undercarriage failure on 10 September 1993 at Leeming.

-**XW525** (S.2B) crashed on 27 July 1977 after losing its tail section following a near miss at Claerwen Reservoir.

-**XW526** (S.2B) crashed on 20 July 1979 near Osnabrück, Germany following fatigue failure.

-**XW531** (S.2B) crashed on 26 November 1974 off the Norwegian coast.

-**XW532** (S.2B) crashed on 25 March 1971 near Venlo, the Netherlands.

-**XW535** (S.2B) crashed on 24 November 1973 near Gutersloh, Germany.

-**XW536** (S.2B) crashed on 16 July 1975 over the North Sea after collision with Buccaneer *XW528* (which landed safely but never flew again!).

-**XW537** (S.2B) crashed on 14 March 1982 on landing approach at RAF Wattisham.

-**XW539** (S.2B) crashed on 4 January 1972 in the Irish Sea near the Isle of Man.

-**XW548** (S.2B) crashed on 3 February 1977 near Volkel, the Netherlands after in-flight fire.

-**XX890** (S.2B) crashed on 18 August 1977 near Laarbruch, Germany.

-**XX891** (S.2B) crashed on 11 August 1983 near Laarbruch, Germany.

-**XX898** (S.2B) crashed on 17 July 1982 on approach of Lossiemouth.

-**XZ430** (S.2B) crashed on 20 May 1984 near Fraserburgh.

Buccaneer S.50 S.A.A.F. :

-**411** crashed after a stall during flight refuelling on 4 January 1973.

-**413** crashed at landing after hydraulic failure on 1 December 1965.

-**415** crashed into the sea at a night flight on 16 October 1969.

-**417** crashed into the sea during delivery flight on 3 November 1965.

-**418** had to be abandoned by the crew after it was damaged during weapon trials on 14 October 1970

-**419** and **420** crashed after a collision near Danger Point on 24 November 1972.

Top view of a Buccaneer S.2B in post-Gulf War scheme. The streaks on the wings are caused by deposit caused by the engine fed Boundary Layer Control system on the wing. The area ruling shape is obvious from nose to tail! (MoD)

Action shot of one of the remaining post Gulf War Buccaneers of no. 12 squadron with the 'Sky Pirates' symbol.
(MoD)

-423   crashed after a double flame out near Scottburgh on 3 August 1978.

-424   crashed on 7 May 1979 near Roedton.

-425   crashed on 18 July 1978 near Lyndenburg.

-426   crashed on 29 December 1975 at Grootfontein after a collision.

It is a long list, but we must realise that the Buccaneer flew mainly at altitudes between 50 and 150 m. If something went wrong there was very little time to take corrective measures and the safest way to survive was to use the ejection seat. The Buccaneers used Cordite explosive cord in the canopy. In an emergency situation the canopy was shattered by the explosion of the Cordite at activation of the Martin-Baker ejection seat. It was a measure just to save precious split-seconds during a bail out of the crew. In most cases the difference between being killed at impact or survival was not more than seconds. We can only conclude that low-flying is a dangerous and risky business!!

Survivors:

In the United Kingdom Buccaneer S.2 *XX885* is being rebuilt to flying condition by Hawker Hunter Aviation. It was granted UK CAA permission to fly in April 2006.
A number of Buccaneers in the UK are in fast taxiing condition and several more are being rebuilt to taxiing condition.
The following complete aircraft are still existing:

XK488   NA.39 Fleet Air Arm Museum (stored), RNAS Yeovilton, Somerset

XK532   S.1 Highland Aviation Museum, Inverness, Scotland

XN923   S.1 Gatwick Aviation Museum, Charlwood, Surrey

XN957   S.1 Fleet Air Arm Museum, RNAS Yeovilton, Somerset

XN964   S.1 Newark Air Museum, Winthorpe Show Ground, Nottinghamshire

XK526   S.2 RAF Honington (gate guard), near Thetford, Suffolk

XN974   S.2 Yorkshire Air Museum (taxiable), Elvington, North Yorkshire

XV361   S.2 Ulster Aviation Society, Long Kesh (stored), Lisburn, Northern Ireland

XN981   S.2B Morris Leslie (stored, dismantled), Errol, Scotland

XT288   S.2B National Museum of Flight, East Fortune, Lothian, Scotland

XV168   S.2B Yorkshire Air Museum, Elvington, Yorkshire

XV333   S.2B Fleet Air Arm Museum, RNAS Yeovilton, Somerset

XV350   S.2B East Midlands Aeropark, East Midlands Airport, Leicestershire

XV359   S.2B Andrew Landon, Topsham, Exeter, Devon

XV863   S.2B Privately owned (for sale), Weston Aerodrome, Dublin, Eire

Technical data S.1 version:

Powerplant:	2 De Havilland Gyron Junior 101 jet engines of 3220 kg (7100 lb) thrust each

Dimensions:
- Length: 19.33 m (63 ft 5 in)
- Wingspan: 12.92 m (42 ft 4 in)
- Height: 4.97 m (6 ft 3 in)
- Wing area: 47.24 m² (508.5 sq ft)

Weights:
- Empty: 13,599 kg (29,980 lb)
- Loaded: 20,412 kg (45,000 lb)

Performances:
Max. speed: 1,038 km/h (645 mph) at sea level
Range: 2,784 km (1730 m)
Service ceiling: 12,200 m (40,000 ft)

Armament
Guns: None
Hardpoints: 4 under-wing pylon stations & 1 internal rotating bomb bay with a total capacity of 3629 kg (8000 lb)

R.A.F. Buccaneer S.2B XV863 'S' in its Gulf War Desert Pink camouflage scheme. It carries the name Sea Witch-Debbie and was sponsored by whisky manufacturer Tamnavoulin

(Srecko Bradic)

XV864 S.2B Defence Fire Training and Development Centre (crash rescue training), Manston Airport, Kent

XV865 S.2B Imperial War Museum, Duxford Airfield, Cambridgeshire

XW530 S.2B Ian Aitkenhead, Buccaneer Service Station, on the A941/North edge of Elgin, Grampian, Scotland

XW544 S.2B The Buccaneer Aviation Group (taxiable), Bruntingthorpe, Leicestershire

XW547 S.2B RAF Museum ('Pauline, Guinness Girl'), Hendon, London

XX885 (*G-HHAA*) S.2B Hawker Hunter Aviation Ltd. (restoration to fly), RAF Scampton, Lincolnshire

XX889 S.2B Gary Spoors, Bruntingthorpe, Leicestershire

XX894 S.2B The Buccaneer Aviation Group, Bruntingthorpe Aerodrome, Leicestershire

XX897 S.2B Atlantic AirVenture, Shannon, Clare, Eire

XX900 S.2B British Aviation Heritage (taxiable), Bruntingthorpe, Leicestershire

XX901 S.2B Buccaneer Aircrew Association, Yorkshire Air Museum, Elvington, North Yorkshire

ZU-AVI (ex *XW988*) S.2B Thunder City (airworthy, for sale), Cape Town, South Africa

ZU-BCR (ex *XW987*) S.2B Privately owned (airworthy), Cape Town, South Africa

ZU-NIP (ex *XW986*) S.2B Ian Pringle (airworthy, for sale), Thunder City, Cape Town, South Africa

XV344 S.2C DERA (gate guard in Qinetiq Complex), Farnborough airfield, Hampshire

412 S.50 SAAF Waterkloof (gate guard by living quarters), Pretoria, South Africa

414 S.50 SAAF Museum, SAAF Swartkop, Pretoria, South Africa

416 S.50 SAAF Museum, SAAF Ysterplaat, Cape Town, South Africa

421 S.50 SAAF Museum, SAAF Swartkop, Pretoria, South Africa

422 S.50 National Museum of Military History, Saxonwold, Johannesburg, South Africa

Except for the 36 Buccaneers as listed above there are a number of Buccaneer nose sections, fuselage/wing sections and other components in possession of various public and private owners.
Source: http://www.thunder-and-lightnings.co.uk/buccaneer/

Technical description:

The fuselage of the Buccaneer was designed using Whitcomb's area rule technique, which had the effort of reducing drag while travelling at high subsonic and transonic speeds, and gave rise to the characteristic curvy "Coke bottle" shape of the fuselage. The majority of the airframe and fuselage was machined from solid casting to give great strength to endure the stress of low level operations.

Considerable effort went into ensuring that metal fatigue would not be a limiting factor of the Buccaneer's operational life even under the formidable conditions imposed of continuous low level flight. A large split air brake was built into the tail cone of the aircraft. The hydraulically operated air brake formed two leaves that

Technical data S.2B version:

Powerplant: 2 Rolls-Royce RB.168-1A Spey turbofans of 5035 kg
               (11,100 lb) thrust each

Dimensions:
-Length: 19.33 m (63 ft 5 in)
-Wingspan: 13.41 m (44 ft)
-Height: 4.97 m (6 ft 3 in)
-Wing area: 47.82 m² (514.7 sq ft)

Weights:
-Empty: 14,506 kg (31,980 lb)
-Loaded: 28,000 kg (62,000 lb)

Performances:
Max. speed: 1,074 km/h (667 mph) at 60 m (200 ft)
Range: 3,700 km (2,300 m)
Service ceiling: 12,200 m (40,000 ft)

Armament
Guns: None
Hardpoints: 4 under-wing pylon stations & 1 internal rotating bomb
               bay with a capacity of 5,443 kg (12,000 lb)
Provisions to carry combinations of:
Rockets: 4 Matra rocket pods with 18 SNEB 68 mm rockets each
Missiles: 2 AIM-9 Sidewinders for self-defence or 2 AS-37 Martel
               missiles or 4 Sea Eagle missile
Bombs: Various unguided bombs, Laser-guided bombs, as well as
               the Red Beard or WE.177A tactical nuclear bomb

Other: AN/ALQ-101 ECM protection pod, AN/AVQ-23 Pave Spike
               Laser designator pod, Buddy refuelling pack or Drop tanks
               for extended range/loitering time
Crew: 2 (Pilot and Observer)

could be opened into the air stream to quickly decelerate the aircraft. The style of air brake chosen by Blackburn was highly effective in the dive-attack profile that the Buccaneer was intended to perform, as well as effectively balancing out induced drag from operating the BLC system. It featured a variable incidence tail-plane that could be trimmed to suit the particular requirements of low-speed handling or high-speed flight; the tail-plane had to be high mounted due to the positioning and functionality of the Buccaneer's air brake. The wing design of the Buccaneer was a compromise between two requirements: a low aspect ratio for gust response and high aspect ratio to give good range performance. The relatively small wing was suited to high-speed flight at low altitude; however, a small wing did not generate sufficient lift that was essential for carrier operations. Therefore, BLC (Boundary Layer Control) was used upon both the wing and horizontal stabilizer, having the effect of energising and smoothing the boundary layer airflow, which significantly reduced airflow separation at the back of the wing, and therefore decreased stall speed, and increased effectiveness of trailing edge control surfaces including flaps and ailerons. For use on aircraft carriers the complete nose and tail sections could be swivelled 180°. Reason for this was that the aircraft had to fit in the aircraft carrier's lift.

Group editor	Corrections
Edwin Hoogschagen	Scott Hochstein
Editor	**Graphic design**
Nico Braas	Jantinus Mulder
Author	**Publisher**
Nico Braas	Lanasta

First print, April 2014
ISBN 978-90-8616-168-3
NUR 465

Contact Warplane:
Slenerbrink 206, 7812 HJ Emmen
The Netherlands
Tel. 0031 (0)591 618 747
info@lanasta.eu

Violearo

© Copyright 2014 Lanasta, Emmen

www.lanasta.com

All rights reserved.
All correspondence regarding copyrights,
translation or any other matter can be
directed to: Lanasta, Slenerbrink 206,
7812 HJ Emmen, The Netherlands.

Buccaneer S.2B XV868 in colors of R.A.F. No. 12 sq. It was scrapped at Eglin in May 1992. (Jantinus Mulder slide collection)'

Below: Buccaneer S.2B XV352 of 237 OCU photographed in 1994. This Buccaneer saw action in the Gulf War as 'Tamdhu' with tail code 'U'. (MoD)

References

- Roy Boot, *From Spitfire to Eurofighter - 45 years of combat aircraft design*, Airlife Publishing U.K. (1990)
- Malcolm English, *"Database: Blackburn (Hawker Siddeley) Buccaneer"*, Aeroplane Monthly, Vol. 40, No. 4, April 2012, pp.69–86.
- A.J. Jackson, *Blackburn aircraft since 1909*. Putnam U.K. (1968)
- Tim Laning, *Buccaneer-The story of the last all-British strike aircraft*, Patrick Stephens Ltd, U.K. (1998)
- Derek Wood, *Project Cancelled*, Tri-Service Press, U.K. (revised edition 1990)

Credits:

With special thanks to Mick Gladwin for proofreading the manuscript.
We also thank the following persons for providing photos and illustrations:
Mick Gladwin, Edwin Hoogschagen, Wim Zwakhals, Minze Veenstra, Johan Visschedijk, Jantinus Mulder, Thijs Postma, Martin Smit.